Leadership Development – Professional Skills to Catalyze Your Career

Topical Conference at the 2021 AIChE Spring Meeting and 17th Global Congress on Process Safety

Online
18 - 23 April 2021

ISBN: 978-1-7138-5335-0

Printed from e-media with permission by:

Curran Associates, Inc.
57 Morehouse Lane
Red Hook, NY 12571

Some format issues inherent in the e-media version may also appear in this print version.

Printed with permission by Curran Associates, Inc. (2022)

For permission requests, please contact AIChE
at the address below.

AIChE
120 Wall Street, FL 23
New York, NY 10005-4020

Phone: (800) 242-4363
Fax: (203) 775-5177

www.aiche.org

Additional copies of this publication are available from:

Curran Associates, Inc.
57 Morehouse Lane
Red Hook, NY 12571 USA
Phone: 845-758-0400
Fax: 845-758-2633
Email: curran@proceedings.com
Web: www.proceedings.com

TABLE OF CONTENTS

Networking in a Virtual World 1
Alaina Levine

Management Systems and Culture Important for Sustained Operations 2
Stuart Arm

Carbon Leadership in a Low Carbon World 4
Robert Zeller

Public Speaking for ChemEs: Improve Your Oral Communication Skills 5
Alaina Levine

So You Want to be a Consultant? 6
David Rosenthal

Can Vapor Channeling on Trays Induce Premature Foam-Flood? 7
Henry Z. Kister, Rohit Kumar

Leading by Example – Innovations in Pharmaceutical Development to Position for a Sustainable Future 29
Moiz Diwan

Enabling Innovation in New Modalities Synthesis 30
Celia Cruz

Getting Ahead of the Curve – Emerging Trends and Success Factors in Biopharmaceutical Development 31
Thomas Nikolai

Imposter Syndrome 32
Dennis Hess

MBBR Wastewater Treatment System for Nitrification of Combined Industrial and Municipal Wastewaters under Cold Weather Conditions with Wide Load Fluctuations 33
Siva Angappan, Eugene Townsley, Somnath Basu

Technical Women's Leadership Forum – What are we waiting for? 34
Belma Erdogan-Haug, Susan Woulfe

Allyship Panel Discussion 35
Brian H. Davison, Sipho Ndlela

4th Pillar of the 3M Strategy "People and Culture" 36
N/A

Case Study: EDI in Practice 37
Jana Green

Great Engineering Skills and High Emotional Intelligence: A Winning Combination 38
Cristina Thomas, Ross Bass

Enhancing Leadership and Management of Individuals and Teams through Emotional Intelligence 39
Dennis Hess

Best Practices Driving Excellence in Plant Performance 40
Alex Kent

Author Index

620743 Networking in a Virtual World

Abstract:

Networking is the most honorable and valuable endeavor in which you can engage, because it is built on a spirit of generosity. At its core, networking is all about crafting win-win alliances where both parties provide value. You may think that "networking" can only take place in person, but this is a myth! In fact, most networking takes place from afar, and in some cases, the individuals may never even meet in person. In this webinar, we will discuss strategies and tactics for finding new collaborators and building mutually-beneficial partnerships with professionals across the globe (and perhaps on some exoplanets too!). We will utilize various methods to find and reach out to people, and start and nurture those critical conversations which can birth amazing collabs! Let's network!

616879 Management Systems and Culture Important for Sustained Operations

Abstract:

Currently, the Department of Energy (DOE) stores approximately 56 million gallons of radioactive and hazardous waste in 177 underground tanks at the Hanford Site. The waste was generated from production of plutonium for the United States' nuclear weapons program starting in the second World War and finishing in 1990. Approximately 20 million gallons of that waste is in a liquid form (supernate), approximately 10 million gallons is in the form of insoluble sludge materials, and the remainder is in a partially soluble solid form referred to as saltcake. Treatment of the tank waste is planned with the Waste Treatment and Immobilization Plant (WTP) being the principle plant where this will be accomplished. The WTP is currently being constructed and large portions of the plant are already starting up. Waste treatment commissioning and operations are planned to start in the early 2020s.

The start of waste treatment operations will mark a significant event in Hanford history. Not since the 1980s, or at least a generation of engineers, will the Site experience the sustained operations of a multi-functional chemical process plant. This situation makes for an interesting study in transitioning management systems and culture from one supportive of design and construction to one conducive to sustained operations. Making ready for safe operations and then to demonstrably show readiness in Operational Readiness reviews and assessments is a well-defined process in the DOE Environmental Management complex. However, there is little programmatic guidance on how to make ready and demonstrably show readiness for sustained operations.

The development of a Sustained Operations Readiness Program was conducted based on the author's professional experience, input from specific individuals with relevant expertise and the author's review of public information. A Sustained Operations Readiness Program is not intended to replace or in any way diminish the role for Operational Readiness Reviews and Assessments as required by the Department of Energy. Rather, the intent is to build upon the Operational Readiness concept to maximize readiness for sustained operations through commissioning and into operations, and to establish a culture of simultaneous operations excellence in production, safety and environmental management. The work has concluded that the state of sustained operations for a complex processing plant can be objectively characterized by 12 features. These features represent the conducive culture or conditions in which the plant, paper and people interact to produce a sustained operations state. The 12 features were further consolidated into the five programmatic and executable elements described below with associated recommendations specific for Hanford waste treatment operations:

Risk management
Risks and opportunities should prioritize continuous improvement initiatives and be informed by the design and technical authorities and stakeholder monitors.
For Hanford waste treatment operations, an integrated risk management program led by the DOE is recommended.

Continuous improvement
All levels of the project organization should exhibit commitment to continuous improvement and learning.
At Hanford, the tank farm contractor had introduced the Lean management system as a continuous

improvement program. The program has yielded efficiencies in tank farm operations and waste treatment operations planning and introduced a culture of continuous improvement. A waste treatment operations program-wide continuous improvement system is recommended for Hanford waste treatment operations.

Continuity of design and technical authorities
The authorities are responsible for managing knowledge for its effective use in the other elements. Recommended for Hanford waste treatment operations is a body of technical program managers from the key technology providers chartered to provide the technical authority and a continuing role for the WTP Engineering, Procurement and Construction contractor as the design authority.

Stakeholder monitoring
Active and deliberate monitoring of stakeholder relationships.
Hanford waste treatment operations is characterized by complex and wide-ranging stakeholders at Federal, State and local levels of government and regulation and various public bodies. Centralized and integrated stakeholder monitoring and management is recommended as an organizational element.

Sustained operations integrated business planning
Business planning integrates and balances technical, financial and regulatory objectives. In a commercial venture this is arguably a business plan.
An integrated Hanford waste treatment operations business plan is recommended for the Site to most effectively compete with other sites in the current Federal environment competitive for Congressional funding.

Readiness evaluation for sustained operations is a continuous function that progresses from overseeing establishment of the five elements during design and construction phases to monitoring the output of those elements performing during commissioning and operations. Continuously assessing readiness for sustained operations becomes an important business function for the Department of Energy's Office of River Protection as competition for Federal funds increases with increasing waste treatment operations throughout the Environmental Management complex.

621580 Carbon Leadership in a Low Carbon World

Abstract:

Carbon dioxide levels in the atmosphere have increased from about 275 ppm prior to the industrial revolution to over 415 ppm today. Nature cannot keep up, and there is now over 2 trillion tonnes of CO2 that has accumulated in the atmosphere. This presentation will touch on point source emissions capture but will focus on Direct Air Capture technology coupled with subsurface sequestration (DAC+S). A technology overview and a summary of potential business models will be discussed.

620745 Public speaking for ChemEs: improve your oral communication skills

Abstract:

Whether you are giving a talk for a class, at a (virtual) conference, or to get a job, it is essential that you shine in every way. Learn the essentials of giving an excellent speech every time. Discover what a speech says about you, how to prepare for a speech, how to combat anxiety, and how to show confidence and poise. Speaking, articulation, and projection techniques, interaction with the audience, and specific presentation issues for scientists/engineers and for science-based conferences will be addressed. And of course, since 100% of our oral presentations are via virtual means, we will delve into the logistics and considerations we can keep in mind as we speak via webcam and present via zoom.

616604 So You Want to be a Consultant?

Abstract:

The practice of consulting can be a very rewarding career choice for someone who has paid their dues in the corporate world and is looking to be their own boss. Consultants provide value to clients for many different situations like solving problems, providing objectivity, and supplementing staff. However before the professional decides to go down the consulting road, they must assess their commitment and personality against the attributes needed for the consulting field. Also, they need to dispel the misconceptions they believe about consulting like the speed to getting rich, escaping the corporate culture, having more time on their hands, and winning project from just the forcefulness of their personality.

Once one has performed this assessment, the list of tasks to be performed for entry is long and time consuming. The first steps involve creating the business plan, followed by deciding the legal and accounting issues, and then deciding your virtual presence. Dealing with clients is far different from when the consultant was sitting on the other side of the desk representing the client. Clients present challenges that must be understood and managed.

This paper reflect my journey starting my consulting business as a Reliability Consultant. I had the advantage of 30 years of industrial experience along with 5 years of consulting experience from an engineering firm which taught me the ropes. I offer this journey to others as a case study to use to judge your commitment to a career change when you decide the 9 to 5 has worn out its welcome.

Can Vapor Channeling on Trays Induce Premature Foam-Flood?

By

Henry Z. Kister

Fluor, Aliso Viejo, CA, USA

and

Rohit Kumar,

Bharat Petroleum Corporation Limited - Kochi Refinery, India

Paper presented at the Distillation Topical Conference, AIChE Spring Meeting, Dallas, Texas, April, 2021

Can Vapor Channelling on Trays Induce Premature Foam-Flood?

By Henry Z. Kister, Fluor, Aliso Viejo, CA, USA and Rohit Kumar, Bharat Petroleum Corporation Limited - Kochi Refinery, India

Combining troubleshooting with a hydraulic analysis identified vapor channeling in fixed valve trays as the root cause of premature foam-flooding and instability in a high-pressure amine absorber.

The lean amine flow rates in the high-pressure (HP) amine absorber in the hydrocracker unit in the Kochi refinery were limited by recurring foamover events. To increase liquid handling capacity, the tower was retrayed by a major tray supplier (Fluor was not involved). The retray was quite standard, using high-capacity valve trays with small fixed valves and push valves and increasing downcomer top area. Surprisingly, higher lean amine flow rates could not be achieved without foamovers. Even at the same amine and gas rates, the frequency of the foamover events jumped from twice per month to once per day. 54 days after the new trays began operation, the bottom drain line from the pre-knockout drum plugged, causing pre-condensed lighter hydrocarbons entrainment into the amine system, and severely aggravating the foaming issues. This required a reduction in both the amine and gas flow rates. The problem was eventually resolved by replacing the tower by a random packed tower and unplugging the pre-KOD bottom drain line (1).

Although the problem was solved, the mystery remained. Why did the revamp trays perform worse than the original trays when they should have performed better? The refinery sought answers from the supplier and the supplier's tray-rating software, but neither of them were able to provide any clues.

Aware of Fluor's distillation expertise, the refinery had Fluor join their task force analysing data and performing a "post-mortem". This joint investigation disclosed a phenomenon that to the best of our knowledge had not been previously reported: vapor channelling inducing premature tray foam-flood. This paper presents our analysis and finding.

Process Description

The Vacuum Gas Oil Mild Hydrocracker (VGO MHC) unit of the Kochi Refinery (1.7 MMTPA) treats VGO from the Crude Vacuum Distillation unit to produce low Sulfur VGO feedstock for the FCC (Fluid Catalytic Cracker) unit, together with conversion of excess VGO to low Sulfur diesel. The design maximum feed sulfur in VGO was 1.6% (by weight).

The effluent from the MHC unit consisting of the cracked products, desulfurized VGO, hydrogen sulfide (H_2S), ammonia, and excess hydrogen are separated in separators/flash drums into liquid and gas streams. The gas streams are further purified in Hot & Cold separators to remove heavier ends and in a High Pressure (HP) amine absorber to remove H_2S. The unit process is described in detail with the aid of a flow diagram in a companion article (1).

Fig. 1 shows the HP amine absorber. Feed gas containing less than 1.5% (by vol.) H_2S enters via a bare nozzle into a pre-Knock Out Drum (Pre-KOD). A small amount of liquid knocked out in this drum drains via the bottom 2" line. Gas from the drum ascends into the absorber, entering it via another bare nozzle. The purified lean gas leaving the HP amine absorber goes to a Downstream Compressor Knock Out Drum (D/S KOD) to remove any amine droplets before being routed to the suction of Recycle Gas Compressor. There is a gas bypass that can divert some of the feed gas to the absorber overhead, at the expense of not removing its H_2S. The absorber pressure is 105 kg/cm^{2}g.

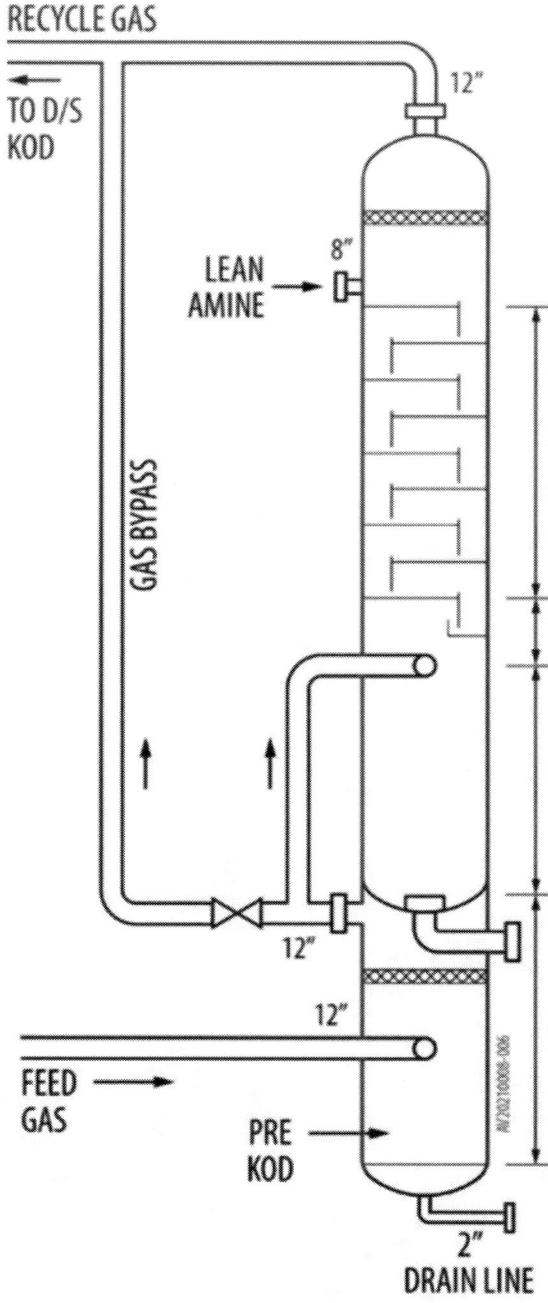

Figure 1. Hydrocracker High-Pressure (HP) Amine Absorber

Vessel inspection during the unit turnaround in November 2017 identified a number of internal cracks and hydrogen blisters in the HP amine absorber. Cracks were eliminated by grinding of the blistered surfaces & weld filling of the internal cracks followed by Post Weld Heat Treatment. In January-March 2019 another unplanned shutdown was taken to partially replace the bottom portion of the absorber shell due to aggravation in the cracks. Details are in Ref. 1.

The major reason identified for hydrogen-induced cracking and hydrogen blisters was high rich amine H_2S loading of more than 0.50 mol/mol of Methyl Diethanol Amine (MDEA), well above the 0.40 mol/mol maximum limit practiced by the industry. The required lean MDEA flow rate in order to maintain the H_2S loading below 0.40 mol/mol of MDEA was limited by the onset of foamover events.

Retray Details and Preliminary Hydraulic Evaluation

The amine absorber is 2.0 m in diameter and had a total of 9-single-pass standard uncaged moving valve trays with sharp (not venturi) orifices at 600 mm spacing. The original trays were designed for 115 TPH of lean amine, but in practice the absorber operated at 70-80 TPH lean amine, with occasional runs just above 100 TPH. To increase the lean amine flow rate, and with the outlook to processing of high sulfur VGO, the tower was retrayed to increase its liquid handling capacity to 150 TPH.

The retray was designed and performed by a major tray supplier with mini-fixed valve trays, with peripheral push valves along the tower walls, pushing the liquid in the direction of flow. To increase the liquid handling capacity, the straight downcomers were replaced by sloped downcomers, with larger downcomer top areas and smaller bottom areas, while keeping the tray active areas unchanged. The downcomer top to bottom area ratio was 2:1, in accordance with good design practices (2, 3). The clearances under the downcomers were also raised to lower downcomer backup. The weir heights were raised to keep the downcomers at a zero static seal. A comparison of the old to the revamp tray design is in Table 1.

The absorber lean amine inlet, rich amine outlet, gas entry, and bottom seal pan remained unchanged. The bottom seal pan overflow weir was 103 mm tall and 1500 mm long, 750 mm below the bottom tray. The straight bottom downcomer was replaced with a sloped one of the same top and bottom areas as the other revamp trays, but its clearance was increased from 50 to 90 mm. No changes were made to the pre-KOD vessel

Table 1. Tray Dimensions

	Old Trays	**New High Capacity Trays**
Tray Type	MR2 & MR63	Fixed Small Valve
Tower area, (m^2)	3.14	3.14
Active area (m^2)	2.43	2.47
Open area (m2)	0.358	0.311
Open area %	14.5%[1]	12.6%
Downcomer Top Area [m^2]	0.354	0.447
Downcomer Bottom Area [m^2]	0.354	0.227
Downcomer Area at top	11.3%	14.2%
Downcomer Top Width [mm]	340	400
Downcomer Bottom Width [mm]	340	250
Flow path length [mm]	1320	1350
Outlet Weir Length [mm]	1503	1600
Weir Height [mm]	50	65
Downcomer Clearance [mm]	40	65
Tray thickness [mm]	2	2

Notes: 1. This is the open slot area of the moving valves

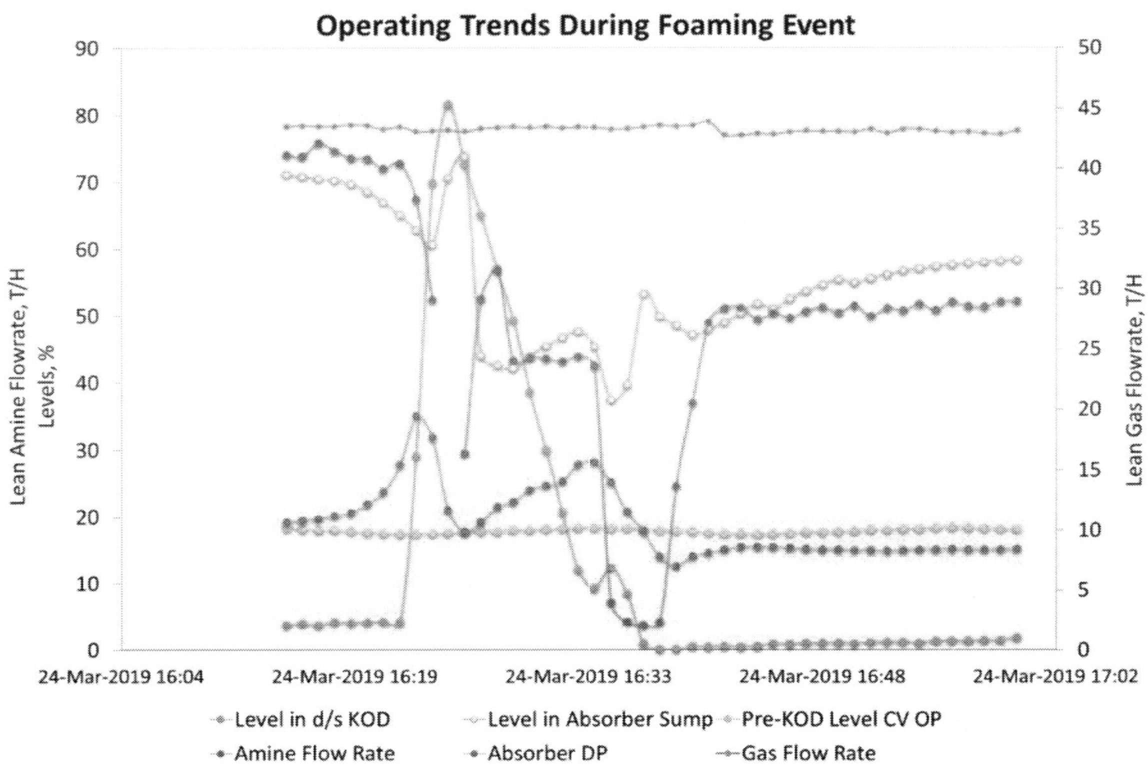

Figure 2. Operating Trends during a Foaming Event, New Trays, before Pre-KOD bottom line plugging

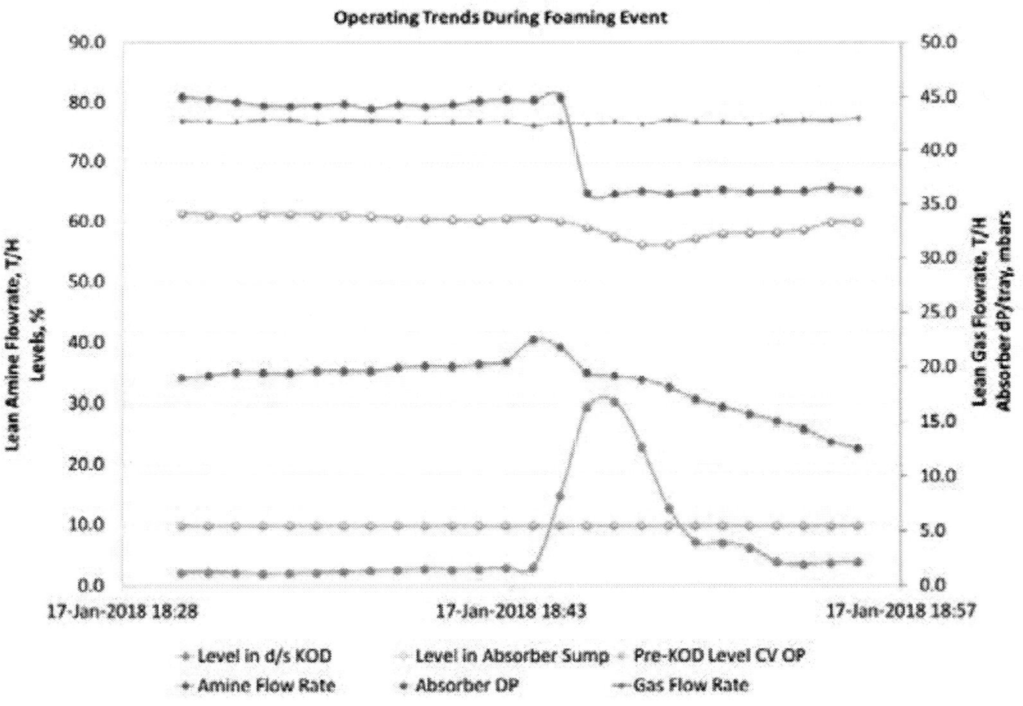

Figure 3. Operating Trends during a Foaming Event, Old Trays

Operating Trends during a Foaming Event

Figure 2 shows the operating trends during a foaming event, a short time after the absorber returned to service following the retray. The tower operated steadily at a gas flow rate of 43 T/H and a lean amine flow rate of 74 T/H, with a differential pressure (dP) of 11 mbar/tray. At about 16:19, the dP started to rise, slowly at first, then faster, peaking at 19 mbar/tray at 16:23. The absorber bottom level fell as the dP rose. Both the rise in dP and the reduction in bottom level are symptoms of flooding. Just before the dP peaked the operators took the corrective action of drastically reducing the lean amine rate to 50 T/H. Amine carryover to the D/S KOD started at 16:23, as seen by the rapid rise in level in the drum. The drum filled up with liquid within a minute or two. Upon reduction in lean amine rate the carryover declined as seen by the more gradual fall in liquid level in the D/S drum, becoming low by 16:37. Simultaneous draining of D/S KOD was carried out at field. Throughout the event, there was little change in the gas flow rate and in the Pre-KOD level.

Figure 3 shows the operating trend during a foaming event, with the old trays. The tower operated steadily at a gas flow rate of 43 T/H and a lean amine flow rate of 80 T/H, much the same as in Figure 2, with a differential pressure (dP) of 19 mbar/tray, higher than in Figure 2. At about 18:43, the dP started to rise, the rise being much less than in Figure 2, peaking at 23 mbar/tray at 18:44. The absorber bottom level slightly fell as the dP rose. Both the rise in dP and the reduction in bottom level are symptoms of flooding. Just as the dP peaked, the operators drastically reduced the lean amine rate to 65 T/H, a higher value than in Figure 2. Amine carryover to the D/S KOD started at 18:45, as seen by the rapid rise in level in the drum. The drum only rose to 31%, much less than in Figure 2, within two minutes. Upon reduction in lean amine rate the carryover declined as seen by the more gradual fall in liquid level in the D/S drum, becoming low by 18:53. Simultaneous draining of D/S KOD was carried out at field. Like in Figure 2, throughout the event, there was little change in the gas flow rate and in the Pre-KOD level.

Figure 3 teaches that following the retray, the events at equivalent operating rates grew in severity and amplitude. In addition, following the retray the foaming events frequency rose from about one-two per month to about one per day. Further, following the retray there were events experienced at as low a lean amine rate of 65 T/H. In summary, following the revamp the foaming events grew in intensity and frequency.

Plugging of the Pre-KOD Drain Line

The Pre-KOD (Figure 1), located upstream of HP Amine absorber, knocks out condensed liquid hydrocarbons to prevent their carryover into the amine absorber. The condensed hydrocarbons leave the drum via a 2" bottom drain line. For the first 54 days after the absorber returned to service following the revamp, there were no signs of plugging of this 2" line, and the pre-KOD drum level was operated and controlled normally. On May 12, 2019, 54 days after the revamp, it was observed during the operation that the drum level control valve became 100 % open throughout the operation. There was no gas blow-by to the downstream cold flash drum from the pre-KOD. This indicated that the 2" bottom line became plugged. De-plugging of this line during plant operation was risky as it was operating under extremely high pressure of 105.0 Kg/cm^{2}g. At a unit shutdown about a year later, the 2" pre-KOD bottoms line was indeed found to be heavily plugged as shown in a photograph in our companion paper (1). Liquid knocked out in the pre-KOD was unable to drain, and would build up until being carried over by the gas stream into the absorber bottom.

Following the plugging of the pre-KOD drain line, the foaming events became far more frequent and far more severe. The lean amine flow rate needed to be reduced from the previous 70-80 T/H to below 40 T/H. The carryover incidents used to occur 10 to 15 times per 8 hour shift when the amine flow was raised beyond 50 TPH. This would annoy the field operators who needed to quickly drain the carried-over amine by opening manual drain valve of the D/S KOD to prevent its high level switch from tripping the lean gas compressor. The repeated carryover events forced keeping an operator full-time at the manual drain valve. The lean amine rate was reduced to 40 T/H, which lowered the frequency of carryover events to 2-3 times in an 8 hour shift at 45 TPH of gas flow rate. But this forced a costly reduction of unit throughput to bring rich amine H_2S loading under control.

It appears that carryover of condensed hydrocarbons into the amine system following the plugging of the 2" pre-KOD drain line caused the severe escalation in foaming events. Condensation of hydrocarbons into an amine solution is a common cause of intensified foaming frequency and severity (4-6).

Troubleshooting and Operation Improvements

Our companion paper (1) lists in detail an extensive troubleshooting and simulation program that shed invaluable light on the absorber operation. Several actions taken to alleviate the severity and frequency of the foaming events were also described there. Steps were taken to avoid hydrocarbon condensation into the amine solution, including keeping the amine sufficiently warmer than the gas to the absorber and knocking out more hydrocarbons in upstream separators by reducing the inlet temperatures. Steps were taken to improve amine quality, including improved filtration, amine washing of the absorber, and discontinuing amine sharing with other units. These were somewhat successful, but the foam events persisted. Noteworthy among these, antifoam injection did not help much and was discontinued after a trial.

The possibility of fouling in the absorber was evaluated. Extensive plugging of the trays was ruled out, as no plugging was observed in the past, the pressure drop was not particularly high, and the carryover problem started immediately following the absorber return to service after the revamp.

The one very successful action was partial opening of gas bypass around the amine absorber (Figure 1), allowing 30 – 40 % of the gas flow to bypass the amine absorber. The gas flow rate through the absorber declined to 25 T/H. This operation brought a huge relief to the plant as the amine carryover was completely eliminated. On the debit side, the unabsorbed H_2S in the absorber bypass necessitated downstream purging out impure recycle gas from the high pressure system to a medium pressure system, where the gas was amine treated and sent to fuel gas. A downside of this measure was loss of H_2.

All these positive changes allowed increasing charge rate to the plant from 140 TPH to 245 TPH, with feed sulfur of 2.2 wt. %. Amine flow was increased up to 125 TPH keeping a H_2S loading of 0.4 moles/mole of MDEA with amine strength of 40%. The continuous filtration of amine also helped the operators to slowly close the bypass over a period of time without any events of amine carryover.

There is one more important operating data point. Most of the time, the unit operated on VGO feedstock. However, from time to time it was switched to diesel feedstock. In the full diesel feedstock operating mode, no carryover was ever observed from the absorber, either with the old or with the new trays, even at 140 TPH of amine flow to the absorber with the gas bypass line fully closed. In this mode the gas rate was 30-32 T/H.

Although gas bypass provided relief to the operators and helped raise the charge rates, the riddle of the intensified amine carryover events remained unsolved. In the quest to explore the reasons, a steady state simulation was developed and provided a basis for hydraulic evaluation. This simulation is described in detail in our companion paper (1).

Initial Hydraulic Evaluation and Problem Fix-up

An initial hydraulic evaluation by the refinery using the tray supplier software failed to uncover any smoking guns. The evaluation showed that with a foaming factor of 0.6 and without gas bypass (gas flow of 45 TPH and amine flow of 100 TPH), the froth heights on trays increased to 76% of tray spacing. This arguably suggests the initial stages of foaming in the absorber. However, a foaming factor of 0.6 is lower than the 0.73-0.8 usually used for these systems (3), and even with this low 0.6 factor, only reached a froth height 76% of the tray spacing, which is still low to support a flood diagnosis. The tray supplier was consulted, but did not have an explanation to offer.

To explain the events at the lower liquid rate (40-50 T/H lean amine flow rate), there was a theory that downcomers may have lost their seal, with downcomer unsealing flood causing the carryover, a mechanism similar to that described by Kister (7) for another amine absorber. Unfortunately the tray supplier's software was unable to determine whether downcomer unsealing was likely, and if so, whether it would lead to flood.

The carryover events were finally eliminated by replacement of the amine absorber with a packed tower (with random packing) and by unplugging the drain line of the pre-KOD. In addition, charcoal filters in the amine regeneration unit were replaced. The issue of amine foaming subsided and the unit has been operating at full throughput ever since. The tremendous savings due to charge increment for just three months is huge as described in detail in our companion article (1).

Rigorous Hydraulic Evaluation and Post Mortem

Although the problem was solved, the refinery continued to investigate the yet unsolved mystery. Aware of Fluor's distillation expertise, the refinery had Fluor join their task force analysing data and performing a "post-mortem".

A review of Table 1 revealed nothing unusual. Increasing the downcomer top area and downcomer clearances are common debottlenecking techniques applied in foaming service, and should have helped to alleviate the foaming events. Even if unsuccessful, they should not have brought the capacity down.

Table 2 identifies the key cases analysed for diagnosing the problem. Any mechanism postulated to explain the absorber mystery must be consistent with the plant experience in all these five key cases.

The flow rates were those measured at the start of the various events, prior to the pressure drop rise and the carryover. The physical properties were per the Refinery's simulation.

The two left hand columns in Table 2 reflect typical normal operation with the pre-KOD operating normally (drain line not plugged). These provide direct comparison of the performance of the old and new trays, unaffected by the entrainment of condensed hydrocarbons into the system that occurred once the pre-KOD 2" drain line plugged. The flow rates are much the same for these columns. Comparing these two cases is central to understanding how the changes in tray design affected the absorber performance. It is important to note that going from the old trays to the new trays the frequency of events rose from 1-2/month to 1/day, and also increased in severity.

The third column in Table 2 reflects the performance of the trays under the severe foaming conditions generated by carryover of heavy hydrocarbons from the pre-KOD into the amine system following the plugging of the 2" drain line. Both gas and liquid rates are considerably lower when the events occurred, and the events were far more frequent. The fourth column reflects the absorber performance with the gas bypass partly open. The gas rate through the amine absorber was quite low. At this gas rate, no foaming events were experienced. The lean amine flowrate could be largely increased with no carryover. Finally, the last column is the diesel feedstock run. Under these conditions, no carryover events were experienced, even at very high lean amine rates.

Table 2 Key Cases Considered in Absorber Hydraulic Analysis

	Old Trays 1-17-2018 Event	Revamp Trays 3-24-2019 Event	Revamp Trays 11-24-2019 Event	Revamp Gas Bypass Partly Open	New Trays Diesel Mode Operation
Frequency of Events	1-2/month	1/day	Many/shift	None	None
Gas Flow Rate, T/H	43	44	30	25	31
Gas Density, kg/m^3	16.3	16.3	16.3	16.3	9.77
Gas Volumetric Flow Rate, m^3/h	2630	2682	1825	1529	3201
Liquid Flow Rate, T/H	75	73	50	100	140
Liquid Density, kg/m^3	1028	1028	1028	1028	1028
Liquid Flow Rate, m^3/h	72.9	71.0	48.6	97.3	136.2
Liquid viscosity, cP	2.88	2.88	2.88	2.88	2.88
Liquid surface tension, mN/m	48.3	48.3	48.3	48.3	48.3

Flooding checks Table 3 provides a flood evaluation for all five cases. The top row expresses the gas load using the C-factor based on the bubbling (active) area, defined as

$$C_b = U_b \sqrt{\frac{\rho_G}{\rho_L - \rho_G}} \quad (1)$$

where C_b is the capacity factor, m/s; U_b is the gas superficial velocity based on the bubbling area, m/s; ρ_G and ρ_L are the gas and liquid densities, kg/m^3.

The C-factors are low, compared to typical values at jet flood for trays spaced 600 mm apart about 0.13 m/s for non-foaming systems and possibly slightly less for foaming systems. The absorber C-factors are not even close. This is echoed by the jet flood calculations, none of which exceeds 40% of flood for cases where events were experienced. Even allowing for a system factor of 0.6, much lower than the values 0.73-0.8 cited for amine systems (3), the percent jet flood will be below 67%. We conclude that the carryover in the absorber has nothing to do with approaching jet flood.

The downcomer backups are low, none exceeding 30.5%, for the cases in which the carryover events were experienced, and for the revamp trays it was 27.1% or less. Even when applying the conservative aeration factor for foaming systems of 0.3 (3), the froth in the downcomer would be only 90% of the tray spacing. Further, Table 3 shows that the new trays would have reduced the downcomer backup, thus debottlenecking a downcomer backup issue. We conclude that downcomer backup flood was not the issue in the absorber.

The downcomer choke limit can be evaluated by considering the downcomer liquid inlet velocities and the apparent downcomer residence times. All the downcomer residence times for the cases in which the carryover events were experienced exceeded 11 seconds, well above the recommended 7 seconds minimum for highly foaming systems (3). We conclude that insufficient downcomer residence time was not the issue in the absorber.

The one parameter that is close to a limit for the cases in which the carryover events were experienced is the downcomer inlet velocity. The recommended value for highly foaming recirculating systems (like amine absorber-regenerator systems) is 0.061-0.076 m/s (3). For once-through foaming systems, the design values recommended are lower, 0.031-0.046 m/s (3). The higher values recommended for recirculating systems is because foaming can be at least partially controlled by antifoam addition. This is usually difficult with once-through system as the antifoam ends up in the bottoms and contaminates the product. Although the current absorber is part of a recirculating amine system, no antifoam was injected, so it will behave like the once-through systems. Our experience with once-through foaming systems, including amine systems with no antifoam injection, is that downcomer choke flood occurs at downcomer inlet velocities of 0.05 – 0.055 m/s. This limit is slightly exceeded in the old trays. We can say with confidence that downcomer choke flooding due to excessive downcomer inlet velocity was the cause of the carryover events in the old trays.

It appears that the tray supplier was concerned about this issue and successfully addressed it in the revamp by increasing the downcomer inlet area. The velocity at the downcomer inlets of the new trays during the foaming events was well below the limit of 0.05 m/s. So, downcomer choke due to excessive inlet velocities is unable to explain the events in the revamp trays.

Another unexplained issue is why the flooding event was not experienced when gas to the absorber was partly bypassed, even though the downcomer inlet velocity exceeded the values at which downcomer choke flood had been experienced.

Table 3 shows that both the maximum downcomer inlet velocity criterion and the minimum downcomer residence time for foaming systems were violated in the diesel run. With a foaming system, this operation should have experienced frequent carryover events. The fact that it did not experience any strongly supports the Refinery's theory that no foaming occurred under the diesel run conditions. A major difference between the diesel and the VGO modes is the cracking of heavy hydrocarbons into lighter hydrocarbons, which is far higher in the VGO mode. Due to intrinsic properties of the catalyst, the refinery has experienced almost negligible cracking when operating in the diesel mode. In contrast, during the VGO run, the catalyst produced a large quantity of these lighter hydrocarbons which got into the amine solution and made it foamy.

The experience of foaming taking place with one campaign but not with another is quite common. Reference 8 reports of 15 cases in which foaming in refinery FCC secondary absorbers occurred when Light Cycle Oil (LCO) was used as the lean oil, and replacing the LCO by naphtha effectively eliminated the foaming. A recent chemical experience (9) tells a story of severe foaming when "substance X", which was heavier than the extractive distillation solvent, was present, but no foaming when it was absent.

In summary, the flood analysis invalidated any theories regarding jet flood or downcomer backup flood or insufficient downcomer residence time having any relevance to the observed carryover events. The flood analysis explained the event experience with the old trays as downcomer choke flood when the downcomer inlet velocity exceeded a limit. The flood analysis also explained the absence of events when operating in the diesel mode by this mode being non-foaming.

The flood analysis was unable to explain the events experienced with the revamp trays. This will be discussed below.

Table 3 Flood Hydraulic Evaluation

	Old Trays 1-17-2018 Event	Revamp Trays 3-24-2019 Event	Revamp Trays 11-24-2019 Event	Revamp Gas Bypass Partly Open	New Trays Diesel
Capacity Factor, C_b, m/s	0.038	0.038	0.026	0.022	0.035
Outlet weir loading[1], m^3/h m	48	34	10	23	65
% Jet Flood[2], FRI	40.0%	35.3%	20.1%	21.0%	38.1%
% Jet Flood[2]	39.6%	39.6%	27.0%	31.3%	46.6%
% Liquid Flow Rate, T/H	75	73	50	100	140
Liquid Height in DC[3], mm	198	180	134	157	238
% Liquid Height in DC[3]	30.5%	27.1%	20.1%	23.6%	35.8%
Liquid Velocity at DC inlet, m/s	0.057	0.043	0.030	0.060	0.085
DC Apparent Residence Time, s	11.4	11.3	16.6	8.3	5.9

Notes
1. Only liquid flowing over the outlet weir, does not include weeping
2. No system factor included
3. Liquid head in downcomer, does not account for aeration

DC unsealing checks: There were theories of gas breaking into the downcomers due to the zero static seals of the revamp trays. The low weir loads during the 11-24-2019 event and during the gas bypass operation suggested possible operation in the spray regime, at which gas can easily break into the downcomers. Once in the downcomer, this gas can prevent the descent of liquid, causing a downcomer unsealing flood in a similar manner to an experience reported in the literature for another amine absorber (7). There was nothing in the tray supplier software to permit this evaluation, but this theory was evaluated during the current rigorous analysis.

Table 4 shows that despite the low weir loads, operation of the absorber was in the froth, not spray regime, based on published criteria (10). The very low C-factors in the absorber render the gas momentum too low to atomize even the small quantity of liquid on the trays. Further, a calculation performed based on the assumption that gas does break the seal into the downcomer showed that its velocity in the downcomer would have been too low to interfere with the liquid descent. This study invalidated the downcomer unsealing theories causing the event at the low weir loads.

Pressure drop and weep checks: The middle part of Table 4 evaluates the pressure drops and weep. While the total pressure drop of the new trays is within the range of common practice, the dry pressure drops of the new trays are very low. A common design practice is to keep the dry pressure drop above 13 mm of liquid. The dry pressure drops of the new trays fall short of this criterion, and are less than half of the pressure drops of the old trays.

At these low pressure drops, weeping is likely from the new trays that would not have occurred with the old moving valve trays. At low gas velocities, the moving valve floats of the old trays would have seated and covered the tray openings, which would have mitigated weeping and kept up the dry pressure drop.

Two correlations have been recommended for weep calculations (10, 11): the Lockett and Banik (12) and the Hsieh and McNulty correlations (13), both of which can be found in accessible references (3, 10). Based on analysis of FRI tray efficiency data, Colwell and O'Bara expressed preference to the Lockett and Banik correlation for low-pressure systems and for the Hsieh and McNulty correlation for high-pressure systems. For the current high-pressure absorber, we preferred the Lockett and Banik correlation, because the physical properties of the liquid and to a lesser degree also of the gas lined up better with the low pressure systems than the high pressure systems that Colwell and O'Bara studied. Therefore our weep calculations are based on the Lockett and Banik correlation, but we also used the Hsieh and McNulty correlation as a check. Table 4 shows that the values calculated from the two correlations were in quite a good agreement.

Both correlations showed a significant amount of weeping from the revamp trays even at the higher gas rates, and much more extensive weep, as much as 50-60% of the liquid, in the cases after the rates were reduced following the pre-KOD bottom line plugging.

Summers et. al. (15) proposed the tray stability factor as a measure of tray stability

$$\eta = (\Delta P_D / H_{CL})^{0.5} \qquad (2)$$

where η is the tray stability factor, dimensionless; ΔP_D is the dry pressure drop, mm liquid; and H_{CL} is the clear liquid height, which can be calculated from the difference between the total and dry pressure drops or from the Colwell equation (14). Data analysis by Summers et. al. show that when the stability factor exceeds 0.6, weeping is less than 30% and the tray is stable. Weeping rapidly escalates at lower stability factors, with their data analysis showing as much as 50% weep or more by the time the stability factor declines to 0.4. Table 4 shows that according to the Summers et. al. criterion, the old absorber trays had stable operation (stability factor of 0.7), but the new revamp trays were highly unstable, operating at stability factors of 0.4 or less.

In summary, the pressure drop evaluation revealed a major issue with the revamp trays: having a very low dry pressure drop and a very low stability factor, both well below the range of good operating practice.

Table 4 Other Hydraulic Parameters Evaluation

	Old Trays 1-17-2018 Event	Revamp Trays 3-24-2019 Event	Revamp Trays 11-24-2019 Event	Revamp Gas Bypass Partly Open	New Trays Diesel
Downcomer Unsealing Check					
Downcomer static seal, mm	10	0	0	0	0
Outlet weir loading[1], $m^3/h\ m$	48	34	10	23	65
Head loss under the DC, mm	20	5	0.5	2.4	19
Flow regime	Froth	Froth	Froth	Froth	Froth
% DC Unsealing Flood[2]	ND[3]	ND[3]	41%	42%	ND[3]
Pressure Drop and Weep Check					
Pressure drop, mm liquid	76	70	55	62	84
Dry pressure drop, mm liquid	25	9.7	4.6	3.0	8.4
Clear liquid height, mm[4]	ND[3]	60	50	59	76
Clear liquid height, Colwell, mm	ND[3]	49	38	53	68
% Weep[5], Lockett and Banik	ND[3]	24.2%	66.1%	61.4%	23.1%
% Weep[5], Hsieh & McNulty	ND[3]	25.5%	45.2%	51.1%	ND[3]
Stability Factor	0.70	0.40	0.30	0.23	0.33
Inlet Vapor Head Check					
$\rho_G V^2$at Inlet, $kg\ m^{-1}\ s^{-2}$	2590	2700	1260	880	2290
Inlet Vapor Head, mm liquid	132	138	64	45	177
Inlet Gas Head/ Dry Pressure Drop	5.3	14.1	14.1	14.1	14.1

Notes

1. Only liquid flowing over the outlet weir, does not include weeping
2. Assuming spray regime and that the downcomer seal is broken
3. ND = Not Determined
4. Difference between pressure drop and dry pressure drop
5. Percent of tray liquid that weeps

Channelling-Induced Foam Flood

Summers et. al. (15) also state that in multi-pass trays, with low stability factors, there is an additional opportunity for the trays to become unstable, totally unrelated to weeping. That instability allows vapor to channel through a part of the tray, travelling upwards without flowing through the other tray passes. One of the authors (15) reported an experience with two-pass trays at low stability factors where he could determine visually that the vapor channelled through a single side of a 2-pass tray tower.

Channelling is a common occurrence on trays where the dry pressure drop is low and at the same time a source of vapor channelling exists. This source will induce vapor to preferentially rise through one section of the tray. The dry pressure drop counters it, tending to distribute the vapor evenly. A low dry pressure drop will be ineffective to counter the channelling source, so the vapor will channel.

Channelling leads to premature flooding, excessive weeping, and reduction in tray efficiency. Common sources of channelling include

1. High hydraulic gradients on the tray, with vapor preferentially rising near the tray outlet where the liquid head is lower. The channelling occurs when the dry pressure drop is too low compared to the hydraulic gradient. This is termed vapor cross flow channelling (VCFC), and has much written about (e.g. 16-18).

2. Reflux or feed maldistribution to multi-pass trays. Here the pressure and downcomer head balances can propagate such maldistribution, especially when the dry pressure drop is low (e.g., 19).

3. Vapor maldistribution. With low dry pressure drops such maldistribution propagates from tray to tray and persists through the tower. In one case, overflow of liquid on a chimney tray generated vapor maldistribution and, due to the low dry pressure drop, it propagated right through the tower.

A likely source of gas maldistribution in the absorber is the incoming gas jet. Table 4 shows that while the inlet gas $\rho_G V^2$ (where ρ_G is the gas density, kg/m^3; and V is the velocity at the gas inlet nozzle, m/s) is well within good design practices, the ratio of the inlet velocity head to the dry pressure drop is extremely high, about 14, in the revamp trays. There is very little dry pressure drop to counter and redistribute the incoming gas jet, and it will rise at the tray end opposite to the gas inlet nozzle as shown in Figure 4. The problem is not excessive inlet gas $\rho_G V^2$, but insufficient tray dry pressure drop. In contrast, the ratio of inlet velocity head to dry pressure drop with the old trays was 2.65 times lower, effectively countering the channelling potential. The old trays had moving valves that closed at the low vapor rates, preserving the high dry pressure drop.

The region where the gas jet would rise in the new trays is across from the gas inlet, as shown in Figure 4. It will generate a high gas velocity region at that end, as shown in red. The gas velocity there will be much higher than in the rest of the active area.

Hartman (18) lists interference of tray support trusses as a contributing factor to channeling, and our past experience concurs. In this tower, the trays had four support beams parallel to the liquid flow path (Figure 5), which would be perpendicular to the horizontal component of the gas flow. The 110 mm deep outside beam across from the gas inlet, which occupied 18% of the tray spacing, would act to trap the channeled gas jet in the outside region opposite the gas inlet. When the gas jet hit the tray above, some of it would be deflected horizontally by the tray, but the beam would resist its dispersion towards the inside of the tower. In contrast, the beam depth prior to the revamp was 75 mm, offering less interference with the gas flow.

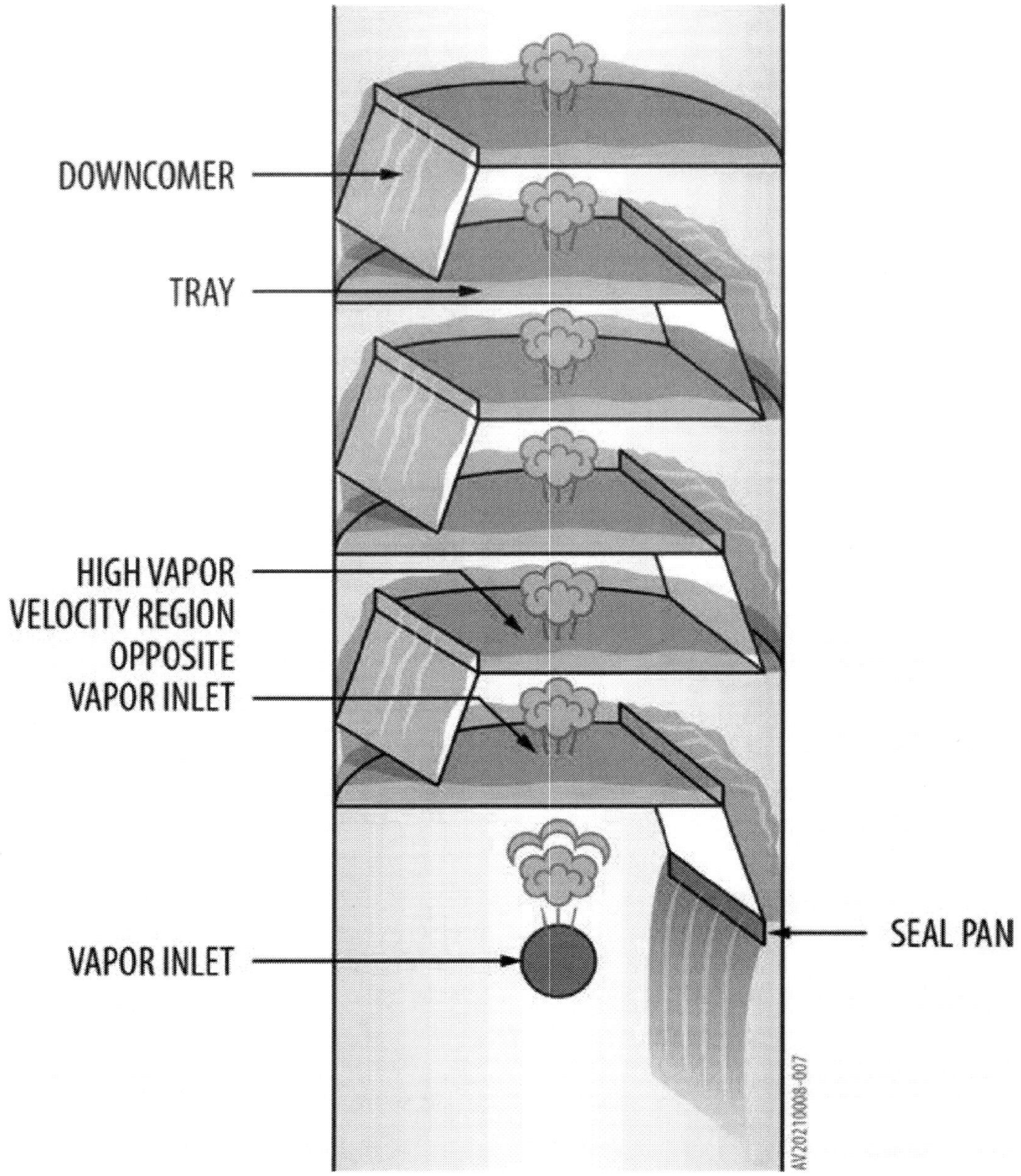

Figure 4. Gas Channeling in Amine Absorber

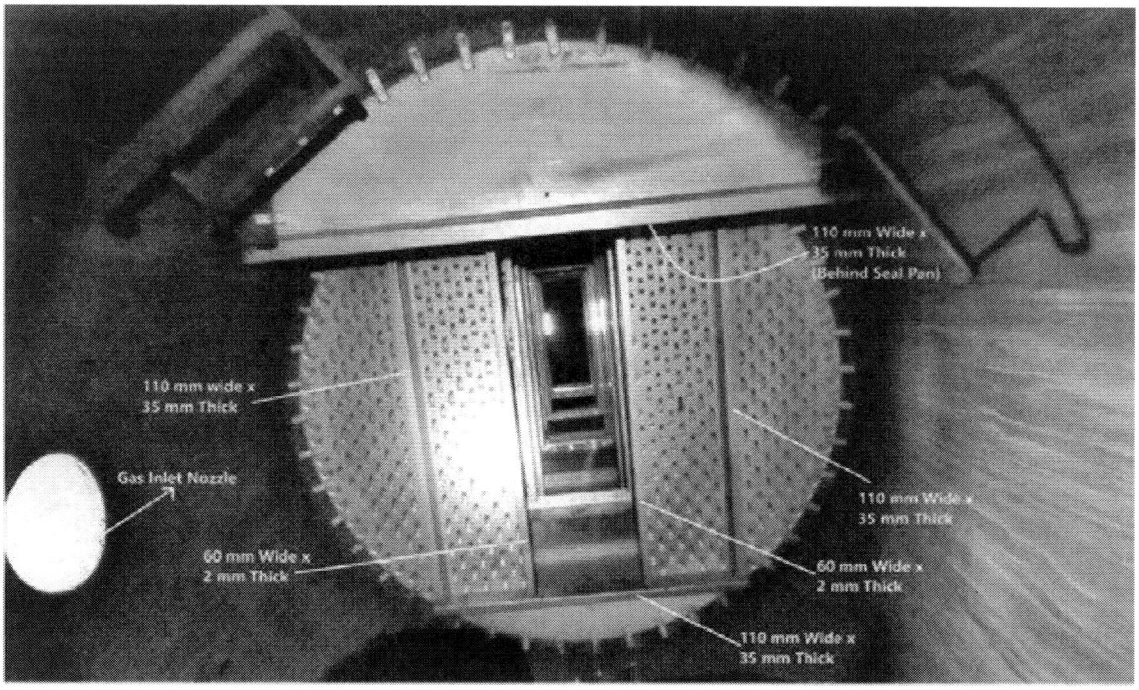

Figure 5 Photograph of the bottom tray with dimensions of trusses marked up (photographer standing in the bottom sump of the tower)

Figure 6 plots data measured by Achoundong et. al. (21) for a foaming chemical system in a test column. It shows that as the gas velocity rises, foaming suddenly takes off at some gas velocity. A similar observation was made by Glausser (22) for a test column using samples from a foaming natural gasoline absorber. Based on these observations, when the velocity of the channelled gas jet through the part of the tray exceeds a certain value, foaming will rapidly escalate and entrain liquid to the tray above. The entrained liquid will increase the downcomer inlet velocity until the downcomer chokes. This action will be augmented by the reduced weeping that is experienced upon foaming.

The combination of the foaming and channelling mechanisms explains why there were premature foam-flooding events like the one on 3-24-2019, even though the calculated downcomer inlet velocity was only 0.043 m/s, and when the foaming was aggravated due to the plugging of the Pre-KOD drain, when the downcomer inlet velocity was as low as 0.03 m/s. It will also explain why the operators successfully got out of flood by cutting back lean amine (liquid). It will also explain why partly opening the gas bypass was successful in mitigating the entrainment events. The gas velocity was lowered to such an extent that the gas jet no longer entrained enough liquid to choke the downcomers. In the diesel mode, the channelling still existed, but the foaming did not, and as a result there was no premature flooding.

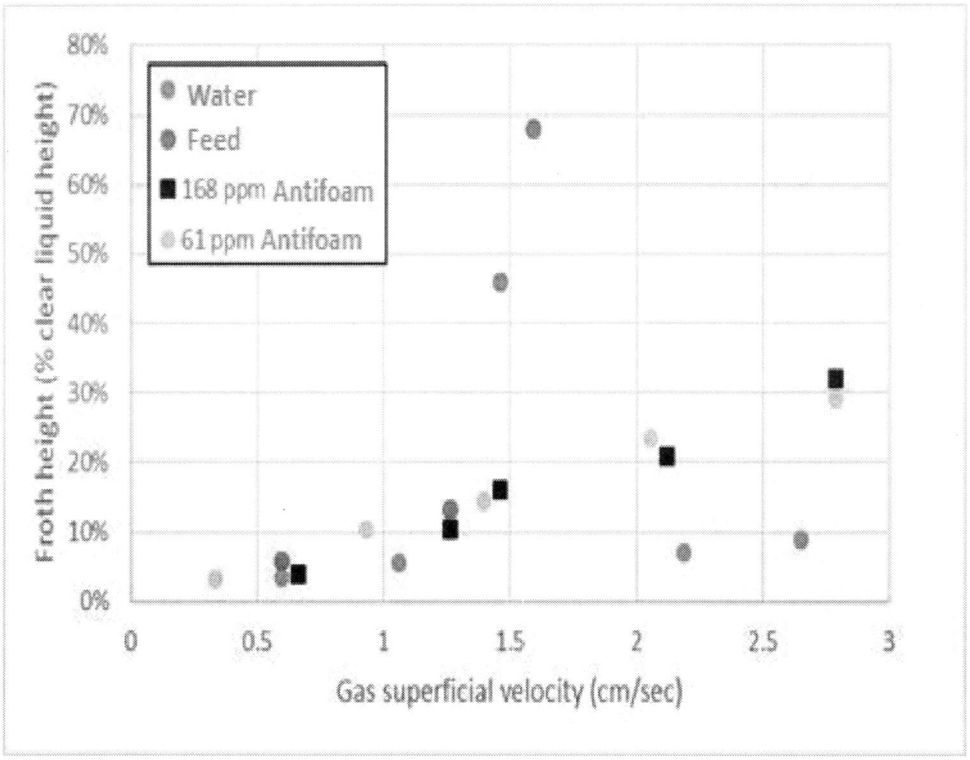

Figure 6. Effect of Gas Velocity on Foam Height. Data obtained in a small-scale test column with a chemical system (from Ref 21. contributed courtesy of Dr. Carine Kuete Achoundong, Dr. Claudio Ribeiro, and Dr. Lihua Liu, BASF).

Finally, one more question. When the tower was replaced by a packed tower of the same dimensions (1), the dry pressure drop would remain low, and the inlet velocity head would remain unchanged. One would expect the channelling to re-occur. The reason why it did not was that the packed tower design added a simple gas distributor at the tower feed inlet (Figure 7), and eliminated the troublesome gas jet.

Epilogue and Takeaways

The capacity of the high-pressure absorber with the old trays was limited by downcomer choke flood at a downcomer inlet liquid velocity of 0.057 m/s, which is the expected foam-flood velocity when no antifoam is applied. At this velocity the absorber experienced 1-2 foam-flood events per month.

The only conceivable reason for the decline is the onset of gas channelling inducing premature foam-flood. The dry pressure drop of the new fixed-valve trays was 2.5 times lower than that of the old moving-valve trays at the operating rates. At the low gas rates, the moving valve floats closed, preventing fall of the dry pressure drop beyond about 25 mm liquid. Not so with the revamp fixed valves, whose dry pressure drops at the same rates fell to about 10 mm liquid, below the minimum 13 mm liquid for good practice. This lowered the tray stability factor from 0.7 (above the minimum recommended value of 0.6) in the moving valve trays to 0.40 in the new revamp trays, signifying high gas channelling potential.

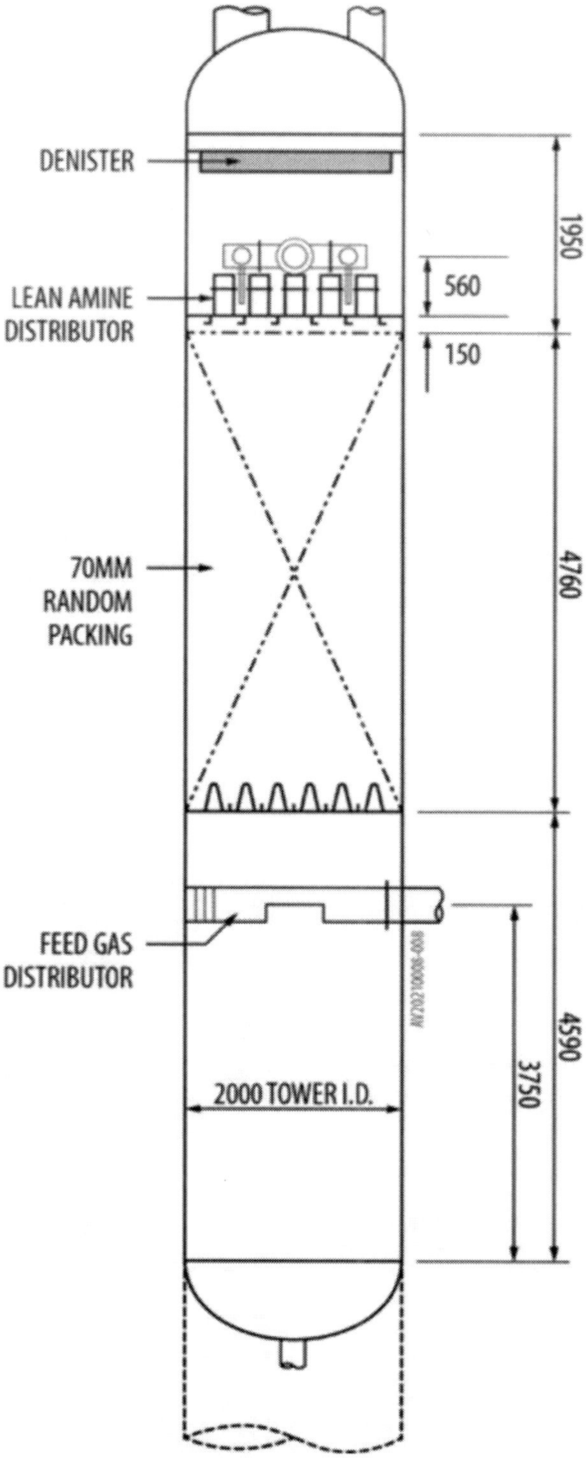

Figure 7. Revised gas entry into the new packed tower. A pipe distributor with a bottom cut-out was added, dispersing the inlet gas jet

Revamp of the absorber with high-capacity fixed valve trays with larger downcomer entrance areas should have permitted higher liquid rates for the same downcomer inlet velocities. Instead, the downcomer entrance velocity at which the foam-flood occurred declined to 0.043 m/s.

The source of gas channelling in the revamp trays was the incoming gas jet. The ratio of gas jet velocity head to the dry pressure drop rose from 5.3 in the old moving valve trays to over 14 in the new fixed valve trays, severely raising the likelihood of channelling. Channelling set in, with high gas velocities in the tray region right opposite the gas inlet. Having 110 mm deep support beams perpendicular to the gas flow may have contributed to the channeling. The foam height at the high-gas-velocity zone induced severe foaming there, which in turn induced entrainment into the tray above. This entrainment overloaded the downcomer inlet velocities, inducing premature flooding in the absorber.

This mechanism, in which gas/vapor channelling induces a premature foam-flood in a foaming service, to the best of our knowledge, was not previously reported. The lesson learnt is that low dry pressure drops should be avoided in foaming systems. Dry pressure drops for foaming systems should exceed 13 mm, and the stability factor (Eq. 2) should exceed 0.6 as recommended (15).

Our analysis also sheds light on the inlet gas velocity from bare nozzles. In the old moving valve trays the ratio of the inlet gas velocity head to the dry bottom tray pressure drop was 5.3, and there was neither apparent channeling nor capacity reduction due to premature foam-flood. When the ratio increased to 14 in the revamp fixed valve trays, gas channelling led to premature foam flood at downcomer inlet velocities at about 25% lower. Interpolating, for a foaming system, keeping the ratio of gas inlet velocity head to the bottom tray dry pressure drop below 6 should prevent channelling from affecting downcomer choke flood. Ratios exceeding about 10 are likely to bring forth channelling and premature downcomer inlet choke floods, the higher the ratio the greater the capacity loss.

Another takeaway is the importance of providing adequate gas distribution when a potential for gas channelling at high ratios of inlet velocity heads to the dry pressure drops exists. This is well-known for packed towers, with guidelines available in the literature (2).

This work also established that a severe foaming environment, such as that created when the bottom drain line from the pre-KOD plugged and condensed hydrocarbons entrained into the system, can severely aggravate the channelled foam capacity loss.

This work highlights the importance of performing a thorough hydraulic analysis to determine the root cause of an operating problem. Running vendor software is a good start, but often does not go far enough and much is left for speculation, which may lead to incorrect diagnosis. A thorough hydraulic analysis is the key to correctly diagnosing tower problems and reaching an effective fix.

Finally, the answer to this article's title question: Can vapor channelling on trays induce premature foam-flood? Absolutely, and we have demonstrated this in this article.

References

1. Mohan, R., R. Kumar, H. S. Kumar Gupta, B. Zohail N., "Amine Anomaly in a Mild Hydrocracker – Learning from Simulation & Field Observations", PTQ, Q1, 2021.

2. Kister, H. Z., "Distillation Operation", McGraw-Hill, NY, 1990.

3. Kister, H. Z., P. M. Mathias, D. E. Steinmeyer, W. R. Penney, V. S. Monical, and J. R. Fair, "Equipment for Distillation, Gas Absorption, Phase Dispersion and Phase Separation", in D. W. Green and R. H. Perry "Perry's Chemical Engineers' Handbook" 9th Ed., Section 14, McGraw-Hill, New York, 2018.

4. Iyengar, J. N., P. W. Sibal, and D.S. Clarke, "Operations and Recovery Improvement via Heavy Hydrocarbon Extraction," 48th Lawrence Reid Gas Conditioning Conference, p. 161, Norman, Oklahoma, March 1-4, 1998.

5. Mykitta, R. S. "Operating Experiences of Shell's Yellowhammer Gas Plant," 49th Laurence Reid Gas Conditioning Conference, Norman, Oklahoma, February 1999.

6. Von Phul, S. A., "Sweetening Process Foaming and Abatement", 51st Laurence Reid Gas Conditioning Conference, Norman, Oklahoma, February 25-28, 2001.

7. Kister, H. Z., and N. M. Mohamed, "Troubleshoot and Solve a Gas Treater Downcomer Unsealing Problem", Part 2, Gas Processing, p.49, March/April 2015.

8. NPRA, Q & A Session on Refining and Petrochemical Technology, 1983, p. 52.

9. Yarbrough, B., T. Cooper, and C. Davidson, "Extractive Tower Column Troubleshooting after Internals Revamp", paper presented at the Distillation Symposium of the AIChE Spring Meeting, Virtual, August 18, 2020.

10. Kister, H.Z., "Distillation Design", McGraw-Hill, NY, 1992.

11. Colwell, C.J., and J. T. O'Bara, Paper presented at the AIChE National Meeting, Houston, Texas, April 1989.

12. Lockett, M. J., and S. Banik, Ind. Eng. Chem. Proc. Des. Dev., 25, p.561, 1986.

13. Hsieh, C. L., and K. J. McNulty, Paper presented at the Annual Meeting of the AIChE, Miami Beach, Florida, November 2-7, 1986.

14. Colwell, C. J., Ind. Eng. Chem. Proc. Des. Dev.,20 (2), p. 298, 1981.

15. Summers, D. R., L. Spiegel, and E. Kolesnikov, "Tray Stability at Low Vapour Load", Distillation and Absorption 2010, Einhoven University of Technology, The Netherlands, p.611, 2010.

16. Kister, H. Z. "Can Vapor Cross Flow Channelling Occur on Valve Trays?" Chem. Engnr. (London), June 24, 1993.

17. Kister, H. Z., K. F. Larson and P. E. Madsen "Vapor Cross Flow Channelling on Sieve Trays: Fact or Myth?" Chem. Eng. Prog., p.86, November 1992.

18. Hartman, E. L., "New Millennium, Old Problems: Vapor Cross Flow Channeling on Valve Trays", in "Distillation 2001: Frontiers in a New Millennium", Proceedings of Topical Conference, p. 108, AIChE Spring National Meeting, Houston, Texas, April 22-26, 2001.

19. Kister, H. Z., B. Clancy-Jundt, and R. Miller, "Troubleshooting a C_3 Splitter Tower, Part 2: Root Cause and Solution", PTQ, p. 39, Q1 2015.

20. Kister, H. Z., K. F. Larson, J. M. Burke, R. J. Callejas, and F. Dunbar "Troubleshooting a Water Quench Tower", Proc. of the 7th Annual Ethylene Producers Conference, Houston, Texas, March 1995.

21. Achoundong, K.,C. Ribeiro, and L. Liu, "Troubleshooting and Redesign of a Solvent Stripper in a Chemical Plant", in Kister Distillation Symposium 2019, p. 605, Topical Conference Proceedings, AIChE Spring Meeting, New Orleans, Louisiana, March 31 – April 4, 2019.

22. Glausser, W. E., "Foaming in a Natural Gasoline Absorber", Chem. Eng. Progr., 50 (10), p.67, 1964.

620242 Leading by Example – Innovations in Pharmaceutical Development to Position for a Sustainable Future

Abstract:

As the active substances and modalities for advanced therapies have become more structurally complex and R&D timelines continue to be compressed, the chemistry and technologies necessary to provide safe, robust and economical access to these molecules needs to keep pace. Enabling technologies such as continuous processing, automation and predictive modeling through first-principles, as well as machine learning can expedite process development, scale-up and technology transfer while ensuring enhanced process efficiency and delivering quality medicines to patients more effectively. The importance of strong leadership and novel concepts in pharmaceutical industry has become even more evident in recent Covid-19 pandemic crisis, different organizations are actively thinking of implementing these paradigms to their benefit. This presentation will share some of the leadership experiences and novel technologies that can be leveraged to accelerate process development and manufacturing and discuss the challenges and opportunities to position for a sustainable future.

620403 Enabling Innovation in New Modalities Synthesis

Abstract:

Dr. Cruz will speak on a subject related to leadership in the pharmaceutical industry.

620244 Getting Ahead of the Curve – Emerging Trends and Success Factors in Biopharmaceutical Development

Abstract:

This talk will discuss leadership aspects involved in pharmaceutical development.

618433 Imposter Syndrome

Abstract:

Imposter syndrome (IS) is a mindset wherein accomplished individuals feel that they have succeeded only due to luck or because others have not realized that they are frauds and thus not deserving of their success. This syndrome is an emotional response rather than a logical one in that the individual dealing with IS knows all details of his/her specific difficulties and struggles and therefore feels that they are undeserving of their achievements. IS is widespread and is characteristic of individuals irrespective of their experience level, gender, ethnicity, stature, and chosen field. IS causes us to avoid applying for certain jobs, negotiating effectively, and enjoying our accomplishments. In this talk, possible reasons that IS is prevalent will be described. Approaches to overcome this syndrome and to minimize the effects of IS on our careers, goals, and subordinates will be discussed

MBBR Wastewater Treatment System for Nitrification of Combined Industrial and Municipal Wastewaters under Cold Weather Conditions with Wide Load Fluctuations

Siva Angappan[1], Eugene Townsley[2] and Somnath Basu[1],

1. Headworks International, 11000 Brittmoore Drive, Houston, TX 77041
2. Batesville Wastewater Treatment Plant, Batesville, AR 72501

Abstract

Uncontrolled discharge of ammonia in treated effluent from wastewater utilities is harmful to the receiving bodies of water. At high concentrations ammonia is toxic to aquatic life. It also adversely impacts the presence of dissolved oxygen which is critical to the health of water bodies. Additionally, discharge of nitrogen in treated wastewater in all forms, including ammonia, promotes undesirable growth of aquatic plants e.g., hyacinth, algae, etc., also called eutrophication, which renders water unsuitable for use and recreational purposes. Environmental regulators throughout North America and globally are increasingly calling for effluent nitrogen control, which requires nitrification as the first step. Nitrification involves biological oxidation of ammonium ions in wastewater. It is kinetically a very slow process and sensitive to environmental conditions, especially wastewater temperature, pH, presence of heavy metals and other toxicants. Many treatment plants, especially in rural US and Canada are lagoon based that can remove only organic pollutants - BOD (biological oxygen demand). In winter, these lagoons are exposed to very low ambient temperatures. Nitrification is difficult to sustain under cold temperatures <5°C. Moving Bed Biofilm Reactors (MBBR) have been shown to effectively conduct nitrification under sustained low temperature conditions while consistently achieving low ammonia limits. Other benefit of the MBBR process include – (a) automatic growth and retraction of the biofilm under varying flows and loads without having the operator to adjust, monitor, or control parameters such as recycled activated sludge (RAS) and mixed-liquor suspended solids (MLSS), (b) requirement of much smaller reactor volume and footprint area, and (c) ability to quickly recover from hydraulic surges, and toxic and shock loads.

Batesville, AR, has a population of around 10,000 but its wastewater treatment plant treats 60,000 population equivalent flows due to many industrial facilities in the area including poultry and dairy industries. Prior to 2015, lagoons were used to treat the wastewater to remove only BOD. Lagoons were also used to settle the solids. Due to substantial growth in industrial discharges to the facility, the plant was stretched to the limit to remove BOD, especially in winter, when the water temperature in the lagoons reach as low as 4° C.

In 2015, the plant's permit was revised to remove ammonia-nitrogen with a year-around limit. Due to the very low water temperatures in winter expansion of the lagoons was not considered a viable option. MBBR was selected as the technology to meet the new regulatory requirements for BOD and ammonia-nitrogen control due to its ability to perform under very low temperatures and handle varying and toxic loads from industrial contributors.

This presentation will consist of an analysis of many years of operating data demonstrating how the MBBR process helped the treatment plant achieve its goals of producing stable effluent quality consistently over the year even under low temperatures and varying loads. Operating performance will be compared with kinetic models based on the principle of MBBR technology.

620741 Technical Women's Leadership Forum – What are we waiting for?

Abstract:

The Women's Leadership Forum (WLF) is one of the largest and oldest employee resource networks (ERN) at 3M. Yet until only recently, relatively few women at the bench (technical track) participated in this ERN. In addition, women reaching upper level technical positions such as Corporate Scientist or Division Staff Scientist, remained disappointingly low over the past several decades. In 2018, female scientists responded by creating a technical chapter (T-WLF) in order to specifically address the needs of technical women. Core WLF pillars of "I believe in Me", "I believe in You", and "We believe in Us" were reinterpreted to meet the needs of technical women. The journey began with the creation of deliberate efforts to 1- *Retain* existing technical employees with Career Development and Technical Men as Advocates committees, 2- Implement best practices and build a better pipeline with *Outreach* committee, and 3- Address the early gap for girls in STEM with *Inspire* committee. In short two years, team has accomplished to collaborate with executive management and amplified the voice of the female technical community, partnered with 3M's Office of Intellectual Property to conduct surveys and focus groups, successfully granted funding from 3M to two non-profit organizations, highlighted successful technical women stories on internal internet, increased presence and engagement with recruiting events, and initiated global expansion. Today, 3M technical women are not waiting; they are part of making change happen.

621920 Allyship Panel Discussion

Abstract:

In the definition from Wikipedia, "*Allyship is the practice of emphasizing social justice, inclusion, and human rights by members of an* ***ingroup****, to advance the interests of an oppressed or marginalized* ***outgroup****. Allyship is part of the anti-oppression or anti-racist conversation, which puts into use social justice theories and ideals.*"

One of the keys to Allyship is communication and the ability to hold difficult conversations. Often, a member of the ingroup wants to be an Ally, but is not sure what would be most supportive and is afraid to ask. After a brief presentation on Allyship, this panel, made up of members of MAC, WIC, DORIC and LBTQ+ will answer questions from the moderator and the audience on how Allies can be the most supportive and effective in both the workplace and life. Please join us at this Panel co-sponsored by the Leadership Development Panel, Management Division, and the EDI Community of AIChE to gain strategies and advice on how to be an Effective Ally.

In 2018, Mike Roman (3M Chairman and CEO) established the 4th pillar of the 3M Strategy, titled "People and Culture" with the strategic goal of strengthening our competitive edge in attracting top talent as well as developing our people & advancing our culture. The three prongs to this pillar include: inclusive workplace, development for all, and future of work (flexibility/global/digital). To continue to build on the strength of 3M's inclusive culture, the CEO Inclusion Council (headed by Mike) was established which includes the Employee Resource Networks' leaders along with Inclusion Leaders in division & staff groups. I was named the Inclusion Leader for the Corporate Research Division in late 2018 and will discuss a variety of Case Studies (best practices and key learnings) from my experience over the last two years.

In my talk, I will cover three different areas: psychological safety, personal advocacy, and organizational leadership. I will discuss my own journey from awareness to allyship to advocacy. I will present case studies including what you can do: mentorship, speaking out for yourself and others, moving beyond your company and into the community as well as driving metrics that facilitate change. Finally, my presentation is meant to help others create an environment where people can show up as their authentic selves and flourish in a diverse, inclusive and equitable environment.

621648 Case Study: EDI in Practice

Abstract:

Part 2 of the session is a case study from a plant manager of DuPont, Jana Green, who took action with her employees after the George Floyd murder, to foster small group discussions. Her advanced EDI practices illustrate how she created psychological safety for the employees to share their feelings, ways she facilitated the group discussions with an approach that was mindful of different stakeholder groups. She exhibited high emotional intelligence throughout the discussions and went on to share them operations leaders across the U.S.

621135 *Great Engineering Skills and High Emotional Intelligence: A Winning Combination*

Abstract:

Every one of us in the field of chemical engineering has invested many hours in the learning and application of principles of *chemistry*, biology, physics, and math to solve problems that involve the production or use of *chemicals*, fuel, drugs, food, and many other products. Indeed, chemical engineers apply mathematics, chemistry and other natural sciences to develop practical ways of using energy and materials for the benefit of society. I love to say we make relevant and impactful chemistry and processes happen. But what about the soft skills? This is where Emotional Intelligence becomes a must have. In this presentation, we will highlight how emotional intelligence or the ability to identify, understand and manage emotions is critical in STEM fields and especially during times of crisis. We will explain how self-awareness and self-regulation, intrinsic motivation, empathy and social skills can help you go further in your career and are must haves for managers and leaders of today.

618409 Enhancing Leadership and Management of Individuals and Teams through Emotional Intelligence

Abstract:

Academic performance and job opportunities in engineering and science positions typically place a premium on intelligence (IQ) and analytical ability. Although these are critical competencies for successful employees and leaders/managers, they are often not the determining factor for success in technical positions. Leaders and managers must be able to understand, interpret, and use their and others' emotions to attain goals and function effectively personally and on teams; this requires emotional intelligence (EI or EQ). Traditional engineering or science programs do not address or even mention EI, although the lack of this skill set is evident during specific course requirements and in various career assignments: whenever teams must function efficiently and effectively. The absence of this trait generally leads to frustration and anxiety for leaders and followers and inhibits many competent and motivated engineers and scientists from pursuing or remaining in leadership/management positions. This talk will describe characteristics of technically-trained individuals who lack EI and the effect that this deficiency has on the individual, on team effectiveness and on the attitudes of subordinates, peers, and superiors. Approaches to improve or develop EI and thereby enhance leadership/management prowess will be discussed.

619090 Best Practices Driving Excellence in Plant Performance

Abstract:

Chemical and Process Plants are becoming increasingly more complex as smart equipment is becoming the new industry norm. Operators, Mechanics and Technical Personnel are constantly inundated with overwhelming streams of data. The growth of plant data has to some extent desensitized our staff to the intended warning flags through this data overload, unfortunately allowing critical information to pass through the system unnoticed. So, despite the sophistication of equipment and monitoring systems, plant failures are still too commonplace. As such, site performance continues to fall short of expectations as evidenced by unacceptable unplanned plant downtime, loss of containment related incidents, operator work-arounds due to equipment unavailability, and high loss profit opportunities. As expected, a retrospective analysis often in the form of event investigations most often reveal that existing plant data and "weak signals" were evident well in advance of the impending failures. Typically, readily available information and precursor events were either not detected, ignored or acted upon. Operating and technical personnel were unable to clearly identify, prioritize, communicate, and mitigate the growing plant risk to prevent the undesired "loss of control" and asset damage.

HOPE Consulting personnel with our broad experience from military, national research labs, commercial nuclear power generation, oil & gas, and mining industries has developed and implemented cross-industry best practices to address these performance shortfalls. The goal of our practices is to support uninterrupted runs between turnarounds, and greatly reduce the likelihood of significant equipment and human failures. These methods assist the operating staff to sort through the live plant data and plant physical anomalies to clearly identify, prioritize and mitigate impending equipment/plant failures. In addition, our process allows your organization to calculate the return on investments (ROI) resulting from our proven methods.

Through the integration of critical plant data, the operating teams can review & analyze performance, identify health threats & adverse trends, prioritize them appropriately and implement timely corrective actions. Team members are trained and become proficient in recognizing these "weak signals" and acting upon them before the equipment conditions become self-revealing e.g. equipment failure (enhancing equipment reliability). Personnel are trained in concept of "building the capacity to fail safe" increasing the likelihood that small failures do not cascade into major losses or incidents. Additionally operations, engineering, and maintenance personnel are mentored in the performance of Asset Integrity Inspections, a lost art as staff become increasingly dependent on process data and plant automation while becoming more disconnected from their plant equipment they maintain and operate.

Through this proven process, work teams have demonstrated their ability to proactively identify degrading conditions and underlying causes and enhanced prevention, rather than reacting to plant events. Additionally, business areas can now more effectively manage the supporting reliability programs and reduce the total number of exceptions and deviations which have contributed to unreliability in the past. Finally, systems and management processes are put in place to facilitate the necessary reinforcement, accountability, and process sustainability.

AUTHOR INDEX

Angappan, Siva....33
Arm, Stuart....2
Bass, Ross....38
Basu, Somnath....33
Cruz, Celia....30
Davison, Brian H. 35
Diwan, Moiz....29
Erdogan-Haug, Belma....34
Green, Jana....37
Hess, Dennis....32, 39
Kent, Alex....40
Kister, Henry Z. 7
Kumar, Rohit....7
Levine, Alaina....1, 5
Ndlela, Sipho....35
Nikolai, Thomas....31
Rosenthal, David....6
Thomas, Cristina....38
Townsley, Eugene....33
Woulfe, Susan....34
Zeller, Robert....4